WITHDRAWAL

Harvest Time Subtraction

Suzanne Barchers

CAPSTONE PRESS
a capstone imprint

First hardcover edition published in 2011 by
Capstone Press
151 Good Counsel Drive, P.O. Box 669, Mankato, MN 56002
www.capstonepub.com

Published in cooperation with Teacher Created Materials. Teacher Created Materials is a copyright owner of the content contained in this title.

 This book was manufactured with paper containing at least 10 percent post-consumer waste.

Editorial Credits
Dona Herweck Rice, editor-in-chief; Lee Aucoin, creative director; Sara Johnson, senior editor; Jamey Acosta, associate editor; Neri Garcia and Gene Bentdahl, designers; Stephanie Reid, photo editor; Rachelle Cracchiolo, M.A. Ed., publisher; Eric Manske, production specialist

Library of Congress Cataloging-in-Publication Data
Barchers, Suzanne I.
  Harvest time subtraction / by Suzanne Barchers.
     p. cm.—(Real world math)
  Includes index.
  ISBN 978-1-4296-6841-5 (library binding)
 1. Subtraction—Miscellanea—Juvenile literature. 2. Harvest festivals—Miscellanea—Juvenile literature. I. Title. II. Series.
  QA115.B355 2011
  513.2′12—dc22   2011001578

Image Credits
BigStockPhoto/Pixelman, 21
Getty Images/Ronnie Kaufman/Larry Hirshowitz, 4
iStockphoto/itographer, 28 (cards); Valarie Staus, 22; ZeNeece, 26
Shutterstock/Alexandar Iotzov, 23; Arena Creative, 28 (baseball player); Ariel
    Bravy, 14; Big Pants Production, 16 (top); Brian K., 12 (pot); Danny E. Hooks,
    17; DaveG, 5; eye-for-photos, 20; Denise Kappa, 25; Gordan Milic, 19; grublee,
    6 (bottom); Gualberto Becerra, 6 (back); Inc, 8; Jane Rix, 10–11; Kameel4u, 6
    (top); Melinda Fawver, 16 (back); Mike Flippo, 18; Olaru Radian-Alexandru,
    24 (front); PeppPic, 12 (note card); Lars Lindblad, 9; Stefano Tiraboschi, cover, 1;
    Stephen Mcsweeny, 16 (bottom); Tatiana Popova, 11 (right); Tatjana Strelkova, 15,
    24 (back); V. J. Matthew, 27

Printed in the United States of America in Stevens Point, Wisconsin.
032011   006111WZF11

Table of Contents

Planning the Harvest Lunch 4
Off to the Store! 16
Time for the Harvest Lunch! 22
Problem-Solving Activity 28
Glossary 30
Index 31
Answer Key 32

Planning the Harvest Lunch

The Garcia kids love the fall. They always go to their grandparents' home for a visit. Their grandparents have a huge apple **orchard**. All the kids and adults help pick apples.

They make a lot of food with the apples. They end the visit with a **harvest** lunch.

There are 5 kids in the Garcia family. This year each kid gets to invite a friend to come to the lunch. It will be a big lunch!

They start planning early in the week. They want to be sure that they prepare enough food.

LET'S EXPLORE MATH

Look at the list. Then answer the questions.

Harvest Lunch Guest List

Adults	Kids	Kids' Friends
Grandma Garcia	Hector	Adrian
Grandpa Garcia	Diego	Lee
Grandma Kane	Eva	Amy
Mr. Garcia	Rosa	Tess
Mrs. Garcia	Maria	Ana
Lee's mother		
Ana's father		

a. How many more kids than adults are on the **guest** list?

b. Grandma Garcia said they can have 20 people for the lunch. How many more people can they invite?

Monday is very busy. The kids pick apples all morning. They love climbing the ladders best.

LET'S EXPLORE MATH

Read the problems below. Use subtraction to solve them.

a. The tallest apple tree in the orchard is 28 feet tall. The shortest is 11 feet tall. How many feet taller is the tallest tree?

b. The tallest ladder is 18 feet tall. The shortest ladder is 10 feet tall. How many feet taller is the tallest ladder?

During their break after lunch they plan the menu. They want to make sure there is enough food. They plan some games too.

Things to Do for Our Harvest Lunch

1. Invite our friends.
2. Check the garden.
3. Decide on the menu.
4. Plan the games.
5. Go shopping.
6. Fix the food.

The kids walk through the garden to find out what is ripe. They make a list of their favorites.

Favorite Foods from the Garden
- corn
- pumpkins
- squash
- tomatoes
- potatoes
- lima beans

Then they look through cookbooks to see what they can make. They need quick and easy **recipes**.

The kids find a great recipe that uses lima beans and corn. Mr. Garcia says that he will make a fire. They can roast corn on the cob and hot dogs.

Succotash

Serves 6

What you need:
- 2 cups fresh lima beans
- 4 cups fresh corn cut from the cob
- 3 tablespoons butter
- $\frac{1}{4}$ cup whipping cream
- $\frac{1}{2}$ teaspoon salt
- $\frac{1}{8}$ teaspoon pepper

What you do:
1. Heat up a big pot of water until it boils. Add a bit of salt.
2. Put in lima beans. Cook about 15 minutes. Drain the water off.
3. Mix up everything else. Put it in the pan with the lima beans.
4. Cook it on low heat until the corn is tender, about 7 to 10 minutes. Stir often.

They will need to buy hot dogs and buns. Mrs. Garcia will make pumpkin bread and dessert.

Menu
- hot dogs and buns
- applesauce
- succotash
- sliced tomatoes
- roasted corn on the cob
- roasted potatoes
- pumpkin bread
- apple pie
- apple cider

LET'S EXPLORE MATH

The recipe on the left makes enough food for 6 people. They will need to make 3 times that much to have enough food for everyone. That will be 18 **servings**.

a. There will be 17 people at the lunch. If they make enough for 18 people, how many extra servings will they have?

b. They picked 39 ears of corn. They need 18 ears of corn for the recipe. How many ears will they have left to roast?

The kids plan their games next. There are 50 apples to use for the games. They decide to bob for apples first. They set aside 24 apples for bobbing. The kids use **subtraction** to figure out that they have 26 apples left over for the other games.

```
  50 apples
− 24 apples
  ─────────
  26 apples
```

Then they will play the apple-toss game. They will see who can toss apples into baskets that are far away.

Off to the Store!

They go to the store on Thursday. They need to buy fresh things like cream and butter. They also need to buy the hot dogs and buns.

They think that each kid might eat 2 hot dogs. The adults will probably only eat 1 hot dog each.

LET'S EXPLORE MATH

Guests	Number of Hot Dogs and Buns	Total Number of Hot Dogs and Buns
10 kids	2 each	20
7 adults	1 each	7

a. If 5 kids eat only 1 hot dog each, how many hot dogs will be left over?

b. If 2 adults do not eat hot dogs, how many hot dogs will be left over?

They buy all the toppings for the hot dogs. Then they pay for the food and go home.

LET'S EXPLORE MATH

Read the problems below. Use subtraction to solve them.

a. Mr. Garcia gave the clerk $45. The grocery bill was $42. How much change did he get?

b. Mr. Garcia came to the store with $69. How much did he have left after spending $42?

One thing they do not need to buy is apples!

To make the apple pies, they peel and cut up 15 **pounds** of apples. For 5 pies they only need 10 pounds of apples. They use subtraction to figure out how many pounds of apples are left over. They will use those apples for a snack.

```
  15 pounds
− 10 pounds
   5 pounds
```

Grandma Kane shows the kids how she can peel a whole apple without breaking the peel!

Time for the Harvest Lunch!

All the guests come right on time. They help make the apple cider. They take turns turning the crank on the apple crusher.

Then they use a **press** to squeeze out the juice. The cider is ready for spices.

The apple-toss game will be the most fun. There are 2 baskets set up. One is for the kids. One is for the adults. The adults' basket is a lot farther away than the kids' basket. The kids use subtraction to figure out that the adults' basket is 6 feet farther away than theirs.

18 feet
− 12 feet
6 feet

Mrs. Garcia does the best! Guess what her prize is? An apple pie!

LET'S EXPLORE MATH

Study the chart. Then answer the questions.

a. Look at Game 2. They moved the kids' basket farther away. Now how much farther away was the adults' basket from the kids' basket?

b. Look at Game 3. They moved the adults' basket farther away. Now how much farther away was the adults' basket from the kids' basket?

Basket	Distance Away Game 1	Distance Away Game 2	Distance Away Game 3
adult basket	18 feet	18 feet	20 feet
kid basket	12 feet	15 feet	15 feet

For lunch they roast hot dogs on sticks. Mr. Garcia pulls the corn out of the fire. Lunch tastes great!

Soon it is time to leave. But Grandpa Garcia has a big surprise for the guests. Each guest gets his or her very own apple tree!

Problem-Solving Activity

Trading Cards with Friends

Don, Taye, and Stefan are friends. They like to collect trading cards. They read the backs of the cards and look at the pictures together. Altogether they have 78 trading cards. Don has 32 cards. Taye has 20 cards. How many cards does Stefan have?

Solve It!

Use the steps below to help you solve the problem.

Step 1: Subtract the number of cards Don has from the total number of cards.

Step 2: Subtract the number of cards Taye has from the number you found in Step 1. That is the number of cards that Stefan has.

Step 3: Add the number of cards Don, Taye, and Stefan have to check your work. They should all add up to 78.

Glossary

guest—a person who is invited to come over to someone's house or party

harvest—the time when crops are gathered

orchard—the place where fruit trees are grown

pound—a unit for measuring weight

press—a machine used to squeeze juice out of fruit

recipe—a set of directions for making food

serving—the amount of food for one person

subtraction—the process of finding the difference between 2 numbers

Index

gardens, 9, 10

guests, 7, 17, 22, 27

harvests, 4–5, 7, 9

orchards, 4, 8

peel, 20–21

pounds, 20

press, 23

recipes, 11, 12, 13

servings, 13

subtraction, 14, 18, 20, 24

ANSWER KEY

Let's Explore Math

Page 7:
a. 3 more kids
b. 3 more people

Page 8:
a. 17 feet taller
b. 8 feet taller

Page 13:
a. 1 extra serving
b. 21 ears left to roast

Page 17:
a. 5 hot dogs
b. 2 hot dogs

Page 18:
a. $3
b. $27

Page 25:
a. 3 feet farther
b. 5 feet farther

Pages 28–29:
Problem-Solving Activity

Stefan has 26 trading cards.